EXAMINING ISSUES THROUGH
POLITICAL CARTOONS

The Environment

Examining Issues Through
POLITICAL CARTOONS

The
Environment

Titles in the Examining Issues Through Political Cartoons series include:

Examining Issues Through Political Cartoons

The Environment

Edited by Laura K. Egendorf

Bruce Glassman, *Vice President*
Bonnie Szumski, *Publisher*
Scott Barbour, *Managing Editor*

GREENHAVEN
PRESS ®

San Diego • Detroit • New York • San Francisco • Cleveland
New Haven, Conn. • Waterville, Maine • London • Munich

LIBRARY OF CONGRESS CATALOGING-IN-PUBLICATION DATA

The environment / Laura K. Egendorf, book editor.
    p. cm. — (Examining issues through political cartoons)
Includes bibliographical references and index.
ISBN 0-7377-1252-X (lib. : alk. paper) — ISBN 0-7377-1251-1 (pbk. : alk. paper)
    1. United States—Environmental conditions—Caricatures and cartoons.
2. Environmental policy—United States—Caricatures and cartoons.
3. Environmentalism—Caricatures and cartoons. 4. Environmental protection—
Caricatures and cartoons. I. Egendorf, Laura K., 1973– . II. Series.

HC110.E5E483 2004
333.7'02'07—dc22

                                          2003066261

Contents

Foreword

Political cartoons, also called editorial cartoons, are drawings that do what editorials do with words—express an opinion about a newsworthy event or person. They typically appear in the opinion pages of newspapers, sometimes in support of that day's written editorial, but more often making their own comment on the day's events. Political cartoons first gained widespread popularity in Great Britain and the United States in the 1800s when engravings and other drawings skewering political figures were fashionable in illustrated newspapers and comic magazines. By the beginning of the 1900s, editorial cartoons were an established feature of daily newspapers. Today, they can be found throughout the globe in newspapers, magazines, and online publications and the Internet.

Art Wood, both a cartoonist and a collector of cartoons, writes in his book *Great Cartoonists and Their Art:*

> Day in and day out the cartoonist mirrors history; he reduces complex facts into understandable and artistic terminology. He is a political commentator and at the same time an artist.

The distillation of ideas into images is what makes political cartoons a valuable resource for studying social and historical topics. Editorial cartoons have a point to express. Analyzing them involves determining both what the cartoon's point is and how it was made.

Sometimes, the point made by the cartoon may be one that the reader disagrees with, or considers offensive. Such cartoons expose readers to new ideas and thereby challenge them to analyze and question their own opinions and assumptions. In some extreme cases, cartoons provide vivid examples of the thoughts that lie behind heinous acts; for example, the cartoons created by the Nazis illustrate the anti-Semitism that led to the mass persecution of Jews.

Examining controversial ideas is but one way the study of political cartoons can enhance and develop critical thinking skills. Another aspect to cartoons is that they can use symbols to make their point quickly. For example, in a cartoon in *Euthanasia*, Chuck Asay depicts supporters of a legal "right to die" by assisted suicide as vultures. Vultures are birds that eat dead and dying animals and are often a symbol of repulsive and cowardly predators who take advantage of those who have met misfortune or are vulnerable. The reader can infer that Asay is expressing his opposition to physician-assisted suicide by suggesting that its supporters are just as loathsome as vultures. Asay thus makes his point through a quick symbolic association.

An important part of critical thinking is examining ideas and arguments in their historical context. Political cartoonists (reasonably) assume that the typical reader of a newspaper's editorial page already has a basic knowledge of current issues and newsworthy people. Understanding and appreciating political cartoons often requires such knowledge, as well as a familiarity with common icons and symbolic figures (such as Uncle Sam's representing the United States). The need for contextual information becomes especially apparent in historical cartoons. For example, although most people know who Adolf Hitler is, a lack of familiarity with other German political figures of the 1930s may create difficulty in fully understanding cartoons about Nazi Germany made in that era.

Providing such contextual information is one important way that Greenhaven's Examining Issues Through Political Cartoons series seeks to make this unique and revealing resource conveniently accessible to students. Each volume presents a representative and diverse collection of political cartoons focusing on a particular current or historical topic. An introductory essay provides a general overview of the subject matter. Each cartoon is then presented with accompanying information including facts about the cartoonist and information and commentary on the cartoon itself. Finally, each volume contains additional informational resources, including listings of books, articles, and websites; an index; and (for historical topics) a chronology of events. Taken together, the contents of each anthology constitute an amusing and informative resource for students of historical and social topics.

Introduction

The American environmental movement has a long history. Since the 1870s John Muir, the Sierra Club, and other important figures and organizations have sought to make Americans aware of ecological issues. However, the U.S. environmental movement has had more than one goal throughout its history. At different points in the past 130 years, the movement has been marked by periods of intense activity, each with a particular type of focus. At first, environmentalists, with the assistance of the government, focused on protecting wild areas through the establishment of national parks. During the 1960s and 1970s, with the national park system long established, environmental activists, Congress, and presidents became more interested in ensuring that America's air and water were clean, as pollution could have serious health effects on people. A second focus at this time was protecting endangered species. Finally, in recent years the dominant debate has been whether businesses, not politicians, are best equipped to be stewards of the environment.

Creating National Parks

In the early years of the American environmental movement many people started to demand that the government protect the nation's wild spaces. This period of environmental activity was marked by a belief in conservationism, which emphasizes the management and development of natural resources. After the Civil War, Americans began to realize that as the population moved westward and established towns and cities, forests and other wilderness areas were vanishing. Prompted by these concerns, Congress passed legislation in 1872 making Yellowstone, located in parts of Wyoming, Idaho, and

Montana, the world's first national park. Four years later the world's first conservation organization, the Audubon Society, was founded. The society, whose initial goal was the protection of birds but which later expanded its interests to include other species and ecosystems, was named in honor of John James Audubon, a naturalist who earned acclaim through his painting of American birds.

While Audubon was an early influence during this period of conservation, much of the success in the creation of national parks can be credited to two men, John Muir and Theodore Roosevelt. Muir was an explorer and naturalist who spent most of the 1870s studying the Yosemite Valley and the Sierra Nevada mountain range. In 1890 Muir's efforts to preserve Yosemite's sequoias (a tree species that can live more than two millennia) led Congress to establish the Yosemite and Sequoia National Parks. Two years later, he cofounded the environmental organization Sierra Club. The mission of the Sierra Club was "to enlist the support and cooperation of the people and government in preserving the forests and other natural features of the Sierra Nevada."[1] Muir's work convinced millions of Americans that the nation's wilderness areas needed to be safeguarded. His views are illustrated in an article he wrote in 1898 for the *Atlantic Monthly*:

> The wildest health and pleasure grounds accessible and available to tourists seeking escape from care and dust and early death are the parks and reservations of the West. . . . The forty million acres of these reserves are in the main unspoiled as yet, though sadly wasted and threatened on their more open margins by the axe and fire of the lumberman and prospector, and by hoofed locusts, which, like the winged ones, devour every leaf within reach, while the shepherds and owners set fires with the intention of making a blade of grass grow in the place of every tree, but with the result of killing both the grass and the trees.[2]

One of the people influenced by Muir was Theodore Roosevelt, who became president in 1901. Roosevelt had been interested in nature since childhood and had majored in natural history in college. In 1887 Roosevelt cofounded the Boone and Crockett Club, an organization that promoted the preservation of animals and their habitats. Roosevelt's and Muir's shared love of nature led to a friendship between the two men, with Muir successfully encouraging the president to expand the national park system, including increasing the size of Yosemite National Park.

National parks were the centerpiece of the first period of intense activity of American environmentalism, as they represented most clearly the belief during that era that large swaths of land needed to be protected from human encroachment. Between 1901 and 1909 Roosevelt signed legislation establishing national parks in Crater Lake, Oregon; Wind Cave, South Dakota; Sullys Hill, North Dakota; Mesa Verde, Colorado; and Platt, Oklahoma. Roosevelt also authorized and made significant use of the Antiquities Act, which was passed by Congress in 1906. As explained by Barry Mackintosh, former bureau historian of the National Park Service, the act authorized "presidents to set aside 'historic and prehistoric structures, and other objects of historic or scientific interest' in federal custody as national monuments."[3] Roosevelt used the act eighteen times during his presidency, including when he proclaimed the Grand Canyon and Arizona's Petrified Forest national monuments.

Roosevelt's support of the environmental movement is evident in a speech he gave in 1907 on the need to conserve natural resources. The president declared:

> We have become great in a material sense because of the lavish use of our resources, and we have just reason to be proud of our growth. But the time has come to inquire seriously what will happen when our forests are gone, . . . when the soils shall have been still further impoverished and washed into the streams, polluting the rivers, denuding the fields, and obstructing navigation. These questions do not relate only to the next century or to the next generation. One distinguishing characteristic of really civilized men is foresight; we have to, as a nation, exercise foresight for this nation in the future; and if we do not exercise that foresight, dark will be the future![4]

While Roosevelt was a strong supporter of the environment, his successor William Howard Taft did not display a similar interest. No major environmental legislation passed during William Howard Taft's administration, but his successor, Woodrow Wilson, continued Roosevelt's work. On August 25, 1916, Wilson signed legislation that created the National Park Service. Operated by the

Interior Department, the service oversaw national parks and monuments. The national parks had been overseen by the Interior Department prior to the legislation but there had not been any unified organization. The law's inception marked the end of the first major period of the U.S. environmental movement.

Protecting Humans and Animals

The environmental movement was mostly in a lull in the five decades after Wilson's presidency. While the federal government did take steps to ensure the continued success of the national park system, and Americans started to worry about air pollution as the nation became increasingly industrialized, the next period of significant activity did not begin until the 1960s. Lasting into the mid-1970s and dominated by Congress, this period was predicated on the belief that the increasingly polluted environment posed a health threat to Americans and that both people and animals were entitled to environmental protections. Such views were expressed in books written by the era's key environmentalists, including Rachel Carson and Paul Ehrlich. During this period Congress passed some of this nation's most powerful environmental legislation, laws dedicated to improving the quality of air and water and preventing the extinction of endangered species.

One of these laws was the National Environmental Policy Act of 1969, which directed federal agencies to ensure that their activities would have a minimal effect on the environment. In 1970 the passage of the Clean Air Act expanded the protection that first began with laws passed in 1955 and 1963 by establishing more stringent standards for outdoor air quality and increasing funds for air pollution research. In 1972 the Water Pollution Control Act (later renamed the Clean Water Act) was passed. Lastly, the Endangered Species Act of 1973 enjoined federal agencies to ensure that any activities they fund or supervise do not place any endangered or threatened animal or plant species in jeopardy or destroy the habitat in which that species lives.

Environmental advocates did not have to rely solely on Congress for help in protecting the nation's people and wildlife. Both presidents Lyndon Johnson and Richard Nixon repeatedly addressed ecological issues in their speeches to Congress. Nixon also established the Environmental Protection Agency (EPA) in 1970, which two years later banned the use of DDT, a pesticide that had been found to be polluting the environment and that was potentially deadly for animals and humans.

One legislator who played an important role in this period of the environmental movement was Wisconsin senator Gaylord Nelson, who founded the annual celebration of the environment, Earth Day. Nelson had spoken on ecological issues throughout the 1960s. In 1969 he decided to organize a grassroots protest that would address the degradation of America's environment. At a conference that September he announced that there would be a nationwide demonstration the following spring. With the help of college students, Senator Nelson's goal was achieved. On April 22, 1970, more than 20 million people participated in the first Earth Day.

The Role of Businesses

In the first and second periods of American environmentalism, the executive and legislative branches of the federal government played important roles in the preservation of the nation's wild areas and in the improvement in air and water quality. With the third and most current period of intense activity, however, debate has begun in earnest as to whether the best stewards for the environment can be found in corporate offices instead of on Capitol Hill. To explain, although America has a long history of national parks, many free market environmentalists contend that privately owned parks are better preserved than those under federal control because private owners have an economic incentive to keep their property thriving and in good shape.

An examination of two regulations that were developed in the 1990s and the first years of the twenty-first century help illustrate this debate over whether businesses should be given greater power over the environment or if their drive for profits will undermine environmental progress. Both policies were modifications of the Clean Air Act. The first policy was a revision to the Clean Air Act that permitted emissions trading. The second set of regulations were the proposed revisions made in 2002 to the New Source Review, or NSR, a program that regulates the pollution emissions created by factories.

Emissions trading began in 1990 as part of a revision of the Clean Air Act. According to the act, the Environmental Protection Agency

(EPA) establishes limits on the amount of pollution that can be emitted by utility plants and other factories. Companies whose level of pollution is below the limit receive credit for the difference. Companies that go over the level—an act that once led to being fined or closed—can purchase credits from "clean" companies.

Supporters of emissions trading contend that the policy encourages companies to develop more environmentally friendly processes because reducing the level of pollutants can help increase their profits. In an article for the journal *Forum for Applied Research and Public Policy*, Kenneth G. Ruffing, who works for the United Nations Department of Economic and Social Affairs, describes the benefits of emissions trading, including the way it can lead to the development of improved pollution control technologies. According to Ruffing, "These technologies . . . can help a company reduce emissions below regulatory limits and then trade its earned credits on the open market. In the process, an entire nation—and, in many instances, the global community—benefits from technological advances."[5]

On the other hand, the trading of pollution credits has been highly criticized by environmental writers and organizations that believe the trading does little to protect the nation's air and water. Environmental activist Brian Tokar explains:

> To true believers in the magic of the free market, it seemed like the perfect plan. But once the EPA actually began auctioning pollution credits in 1993 it became clear that virtually nothing was going according to their projections. The first pollution credits sold for between $122 and $310, significantly less than the agency's estimated minimum price, and by 1995, bids at the EPA's annual auction of sulfur dioxide allowances averaged around $130 per ton of emissions. As the value of the credits declines, the incentive to buy credits rather than invest in pollution controls becomes increasingly attractive. Air quality can continue to decline, as companies in some parts of the country simply buy their way out of having to comply with pollution reductions.[6]

Pollution credits are not the only controversial change in the Clean Air Act. The controversy over another decision to alter long-standing EPA air pollution rules is a further example of the conflict between industrialism and environmentalism that typifies the third period of the environmental movement. In this case the changes to the Clean Air Act targeted the New Source Review. The NSR, included in the Clean Air Act in 1977, required owners of polluting factories to install up-to-date pollution reduction technology whenever changes to the factory caused an increase in pollution output. At the same time, the NSR allowed existing factories to continue normal operations as long as they met the other requirements of the Clean Air Act. In a report on the NSR the Sierra Club states that the goal of the 1977 law was to gradually replace older, more polluting factories with newer and cleaner plants. What happened instead, the Sierra Club notes, is that "older factories . . . are still in business, and are still putting out more pollution than their modern counterparts. The ability to avoid regulation has almost certainly played a part in their continued operation."[7] The organization asserts that failures to comply with the NSR cause as many as seven thousand deaths and one hundred thousand asthma attacks each year.

Efforts to reform the NSR began during the Clinton administration, but no changes were made until after George W. Bush took office. Under the Bush proposal, which was announced on December 31, 2002, and was scheduled to take effect one year later (the proposal was temporarily blocked by a federal court in December 2003), industrial facilities would not need to purchase new antipollution devices if the cost of plant improvements is less than 20 percent of the value of the plant. Additionally, equipment that is no more than fifteen years old would not need to be upgraded when it becomes more polluting if the equipment was considered "clean" when first constructed. The EPA supported these changes, noting in a review that reforming the program would make it easier to modernize facilities and create new technologies.

The decision to alter the New Source Review has worried several prominent environmental organizations. For example, the National Resources Defense Council (NRDC) is concerned about the reforms' ramifications. The NRDC asserts that the revised NSR will limit the ability of state governments and national park managers to control air pollution and protect America's wilderness from smog and acid rain. Furthermore, the council opines that changes to the NSR will make it significantly easier for factories to ignore their responsibilities to the environment.

Most significantly, the debate over the NSR has become a legal matter. On October 27, 2003, thirteen states and more than twenty cities filed lawsuits in an effort to keep the reforms from becoming official. According to these suits, the new regulations weaken protections of human health and the environment. The states and cities further contend that only Congress, not the president, can initiate such a significant change to the Clean Air Act.

Many commentators consider these concerns specious. *Detroit News* columnist Thomas J. Bray suggests that the NSR of the past is at least partly to blame for the rising gas prices that occur when oil supplies are disrupted. In Bray's view, the NSR bureaucracy discouraged the building of new refineries. Bray is among those who believe that altering the NSR will lead to businesses taking the initiative to protect the environment. As Bray contends, "the EPA's change of heart on New Source Review suggests . . . an appreciation that the role of government is to buttress the workings of the marketplace rather than disrupt it. No country ever saw its environment get greener without first getting more productive and thus richer."[8]

Emissions trading and the reforms to the NSR are two examples of the issues facing the people involved in the third major period of the U.S. environmental movement. For America's environment to thrive, it will likely be necessary for industrialists, environmentalists, and politicians to work together to find solutions that are beneficial to all parties. Unless that happens, any proposed environmental reforms are likely to be viewed with suspicion and seen as either attempts to shackle the free market or efforts to benefit polluters' pocketbooks at the expense of human health.

The Debate Continues

For more than a century Americans have sought to protect and preserve their nation's environment. The U.S. environmental movement has changed throughout the years, from the first period of intense activity, when the focus was on the protection of America's wilderness and the establishment of national parks, to the second period, which centered on protecting the health of humans and endangered species, to the third and current phase in which people are examining whether businesses are better qualified than government officials to be environmental stewards. Of course, other issues such as global warming, overpopulation, and rain forest destruction have also been the center of much debate. In *Examining Issues Through Political Cartoons: The Environment*, the cartoonists explore a variety of controversial environmental issues. Their disparate offerings show that while protecting the environment might be a common goal, finding the most effective approach remains a controversial challenge.

Notes

1. Original Sierra Club Mission, quoted in Michael Cohen, "Origins and Early Outings," www.sierraclub.org.

2. John Muir, "The Wild Parks and Forest Reservations of the West," *Atlantic Monthly*, 1898.

3. Barry Mackintosh, "Theodore Roosevelt and the National Park System," National Park Service, 1999. www.nps.gov.

4. Theodore Roosevelt, address to a 1907 governors' conference.

5. Kenneth G. Ruffing, "Achieving Sustainability," *Forum for Applied Research and Public Policy*, Winter 1999.

6. Brian Tokar, "Trading Away the Earth: Pollution Credits and the Perils of 'Free Market Environmentalism,'" *Dollars and Sense*, March/April 1996.

7. Sierra Club, "Background on New Source Review and the Clean Air Act," Sierra Club, August 2002. http://minnesota. sierra club.org/air_toxics_background_on_NSR.htm.

8. Thomas J. Bray, "Down and Dirty at the EPA," *Opinion Journal*, November 26, 2002. www.opinionjournal.com.

Chapter 1

Environmental Crises

EXAMINING ISSUES THROUGH
POLITICAL CARTOONS

Preface

To environmentalists the world is facing myriad problems, from global warming to acid rain. One ecological issue that has received considerable attention is the growth of Earth's human population. Many environmentalists fear that as the population soars, finite resources such as oil and water will become overly strained. However, other analysts contend that this fear is unfounded and that overpopulation will not be a serious problem.

By 2020 the world's population could hit 8 billion, up from its current level of 6.3 billion, and feeding all those new mouths may prove nearly impossible. Environmental writers Paul and Anne Erlich write in their book *Betrayal of Science and Reason: How Anti-Environmental Rhetoric Threatens Our Future* that even the development of a variety of rice that increases production by 10 percent will not be enough to prevent hunger: "A 10 percent increase in rice production may sound impressive, but set against growing human numbers in Asia (where 90 percent of the world's rice is consumed), that increase would be barely sufficient to support five years of population growth."

The rising population has also been linked to problems beyond food shortages. Don Hinrichsen, writing for *International Wildlife*, contends that "human activities caused by population growth and consumption patterns are taking a heavy toll on our planet's life-support systems." Other environmentalists have contended that because of the population boom, worldwide energy use may double by the middle of the century. This spike in energy use would result in an increase in carbon dioxide and other greenhouse gases that reportedly lead to global warming.

Overpopulation could also affect the United States. According to the Carrying Capacity Network (CCN), an organization that seeks ways to stablize America's population, the U.S. population will reach 560 million by 2060, nearly double what it is today. In its publication *Food, Land, Population, and the U.S. Economy*, CCN examines several of the problems that may emerge as the United States becomes more populous. For example, arable cropland is being lost every year to a number of factors, including urbanization and erosion; by 2060, the amount of tillable land will have plummeted 75 percent, from 470 million acres to 120 million acres. According to CCN, "Only 0.6 acres of arable land per person will be available in 2050, whereas more than 1.2 acres per person are needed to provide a [diverse] diet (currently, 1.6 acres of arable land are available)." The organization also contends that almost all of the natural gas and oil in the United States will be gone by 2050 and that by 2060 Americans will have at best the minimum amount of water required for basic human needs, seven hundred gallons per person per day.

Not everyone agrees with these doomsday predictions. Some instead contend that the population will grow less rapidly than predicted and that oil, gas, water, and food will not run out. In his article "Population Growth: Ending the Myth of Overpopulation," Joseph L. Bast, the president of the Heartland Institute, a research institution that aims to empower people and build social movements, cites United Nations' predictions that the world population will peak at 9 billion, a figure likely to be reached by 2050; this prediction runs counter to earlier claims by some experts that the population could climb as high as 12 billion.

Even 9 billion is one-and-a-half times the present-day population, but Bast and other analysts are not worried about the potential consequences. Unlike the Ehrlichs, they believe that crop yields will continue to grow, proof of what they consider the continuing success of the Green Revolution—an agricultural movement that began during the late 1960s, when scientists developed high-yield varieties of wheat and rice that allow farmers to harvest larger crops without requiring additional land. As Bast writes, "Could the world's farmers feed 9 billion people? Easily. . . . The widespread adoption of high-yield farming methods already being used in developed nations would enable farmers to double their output without increasing the number of acres under cultivation."

Bast is also among those who are certain Earth will not run out of oil, gas, or water. He argues that oil reserves are in fact larger than they were three decades ago because extraction techniques have improved. Brian Carnell, the webmaster of www.overpopulation. com—a website that asserts that the planet is not overpopulated because enough food and water are available to "provide everyone on Earth a living standard at the subsistence level"—argues that the water supply will not sink to dangerously low levels. Rather, he maintains, "The major cause of water shortages are governments that artificially lower the price of water."

It could be at least five decades before humanity knows whether it has overextended Earth's natural resources, but that may not be the only environmental crisis of the twenty-first century. In the following chapter the cartoonists consider what, if any, environmental problems exist in modern society.

Examining Cartoon 1:
"Ouch"

About the Cartoon

In this cartoon Golliver offers an overall look at the way he believes humans are destroying the earth. He uses an aphorism by Polish writer Stanislaw Jerzy Lec to express the idea that every act of destruction, no matter how small it seems, has helped bring pain to Earth (depicted here as a pained and wounded face). Among the problems that have harmed the environment, according to Golliver, are drilling in the Arctic, oil spills, air pollution, and over-population. Each of the quotes in the cartoon are common excuses for actions that either harm the environment or fail to protect it;

individually, these statements seem harmless but their cumulative effect is troubling, in Golliver's view. Golliver is suggesting that until people are ready to admit their culpability for the state of the earth, this destruction of the environment will continue.

About the Cartoonist

Golliver is the pen name of Gary Oliver. Oliver is the editorial cartoonist for the *Big Bend Sentinel* in Marfa, Texas.

Examining Cartoon 2:

"Hee! Hee! False Alarm!"

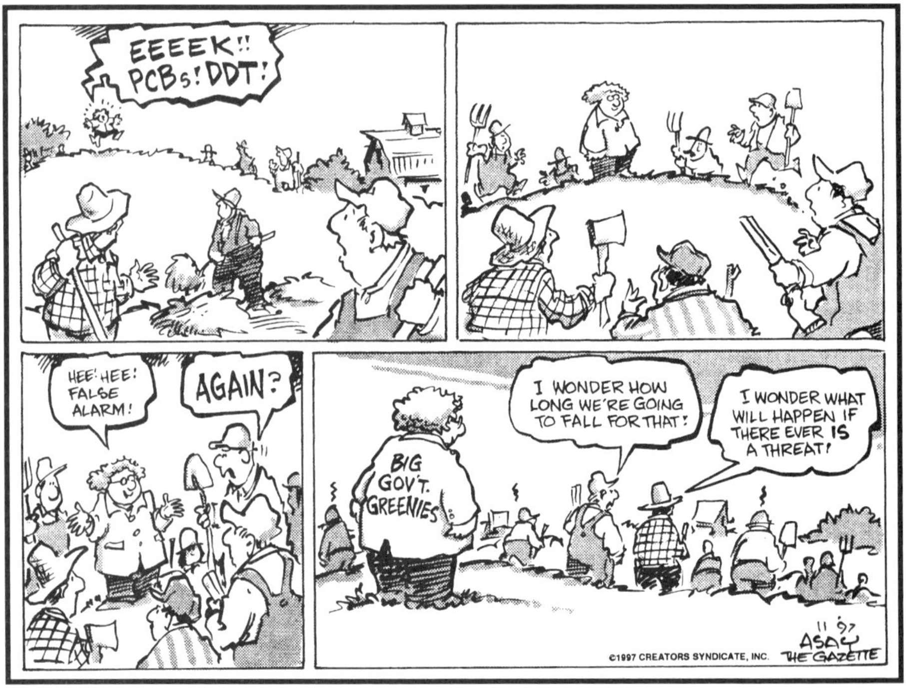
EEEEK!! PCBs! DDT!
HEE! HEE! FALSE ALARM!
AGAIN?
BIG GOV'T. GREENIES
I WONDER HOW LONG WE'RE GOING TO FALL FOR THAT!
I WONDER WHAT WILL HAPPEN IF THERE EVER IS A THREAT!
©1997 CREATORS SYNDICATE, INC.
ASAY THE GAZETTE

About the Cartoon

Chuck Asay asserts in this cartoon that government agencies are like "the boy who cried wolf," exaggerating threats to the environment. Asay targets inaccurate claims made about polychlorinated biphenyls (PCBs) and the pesticide Dichloro-diphenyl-trichlor-oethane (DDT). PCBs are oily liquid or solid mixtures of as many as 209 chlorinated compounds used as lubricants and coolants in electrical equipment. The United States ceased production of

PCBs in 1977 because of evidence that they caused significant damage to the environment and harmful health effects, including cancer and liver damage. However, subsequent studies have found no association between PCB exposure and cancer or other diseases. The Environmental Protection Agency (EPA) banned the use of DDT in 1972 due to the belief, first stated by environmentalist writer Rachel Carson, that the pesticide caused cancer and negatively affected bird reproduction. According to the *British Medical Journal*, no such connection can be found between DDT and cancer patterns. Furthermore, EPA hearings have revealed that DDT cannot be linked to thinning bird shells or the decline in certain bird populations. In his third and fourth panels, Asay suggests that these false alarms could inure people to future claims of environmental problems. As a consequence, a real environmental threat would likely be ignored because Americans would expect it to be yet another false alarm.

About the Cartoonist

Chuck Asay is the editorial cartoonist for the *Colorado Springs Gazette* and is syndicated with Creators Syndicate in more than eighty newspapers nationwide.

Examining Cartoon 3:
"Earth Day with President Bush"

About the Cartoon

In this cartoon Chris Britt criticizes President George W. Bush's policies by depicting him celebrating the environmental holiday Earth Day by finding ways to destroy the environment. Britt targets Bush policies allowing increased digging for coal, drilling for oil, clear-cutting forests, and permitting higher levels of arsenic in drinking water. The first policy has been decried by environmentalists, who assert that coal plants are a major source of air pollution and have been shown to release particles that significantly increase deaths from lung cancer and heart disease. Bush also wants to drill the Arctic National Wildlife Refuge (ANWR) for oil, a move that many people believe will harm the environment and endanger caribou, polar bears, snow geese, and other wildlife. Clearcutting is a logging method in which all the trees in a certain area are removed, the surrounding vegetation is burned, and the land is replanted with one species of trees. As a result, according to environmentalists, fragile and diverse ecosystems are destroyed, water quality worsens, and species become endangered as their habitats are destroyed. Lastly, the Bush administration has raised the acceptable level of arsenic in drinking water, despite concerns raised by the National Academy of the Sciences that this increase will lead to a heightened cancer risk in communities whose water contains arsenic. The new permissible level is ten parts arsenic per billion parts water, more than three times the academy's recommendation of three parts per billion.

About the Cartoonist

Chris Britt is the editorial cartoonist for the *State Journal-Register* in Springfield, Illinois. Britt has won several major awards, including the National Press Foundation's Editorial Cartoonist of the Year in 1994.

Examining Cartoon 4:
"How to Spot an 'Environmentalist'"

HOW TO SPOT AN "ENVIRONMENTALIST"

BOB LANG © 1-13
EDITORIAL SERVICES
RightToons.com

THAT LINGERING ODOR OF (GASP) GOATS AND POT...

COMPLIMENTARY CANOE TIED TO ROOF (NEVER USED)

SUV FOR 'OFFICIAL' ENVIRONMENTAL BUSINESS. (DON'T YOU DARE TO CRITICIZE...)

MANDATORY "PRO-CHOICE" "SAVE THE EARTH" STICKERS.

PEACE AT ANY COST
PRO CHOICE

GORE 04   PETA

EARTH TONE COLORED PAINT SCHEME.

OPTIONAL "GORE '04" BUMPER STICKER.

TRAIL OF (BIODEGRADABLE) TRASH BEHIND VEHICLE.

About the Cartoon

Bob Lang decries the hypocrisy of environmentalists in this cartoon. He characterizes them as making superficial gestures in support of the environment while engaging in environmentally destructive behavior. To illustrate this point, Lang depicts an environmentalist driving an SUV, a vehicle that has been roundly criticized by environmental organizations for using more gasoline than smaller vehicles and thus creating more air pollution and requiring more of the world's limited oil supply. To create an environmentally friendly image, the SUV is painted in earth-tone colors, has a never-used canoe attached to the roof, and is emblazoned with pro-environment stickers. In Lang's opinion environmentalists are pot-smoking litterbugs who are eager to embrace any liberal candidate or cause but are unwilling to face any criticism for their inconsistencies.

About the Cartoonist

Bob Lang is a political cartoonist whose works have appeared in *Insight* magazine, the *Washington Times*, and the *Washington Post*. He also contributes cartoons to several websites, including CNSNews.com and CNN All Politics. On two occasions Lang has won the National Newspaper Publisher's Best Editorial Cartoon of the Year award.

Chapter 2

Economy or Ecology?

Preface

The health of the environment and the health of the economy are sometimes seen as mutually exclusive. For example, the costs of reducing the amount of pollutants released by a factory are often considered prohibitive. In addition, the protection of endangered species often comes at the expense of property owners who are prohibited from developing the portions of their land where those species reside. However, some commentators have suggested that a thriving free-market economy is the best way to protect the environment.

In the view of free-market advocates, giving private entities control of wild spaces is the most reasonable way to keep the environment clean and thriving. They argue that private property rights are more effective than government regulations because private owners have an economic incentive to keep their land pristine. Jane S. Shaw, a senior associate at the Political Economy Research Center, a foundation that researches environmental issues, writes in the anthology *A Blueprint for Environmental Education*, with her coauthor Terry L. Anderson, "Private ownership makes people accountable. People must bear the costs of actions that decrease the value of the resources they use and they can reap the rewards of actions that increase the value of the resources." Shaw and Anderson cite several examples of successful privatization, including New York City's Central and Bryant Parks. While the city continues to own both parks, the day-to-day operations are overseen by private businesses that work to ensure that the parks remain clean and safe. Privately owned and operated parks in less urban areas, such as North Maine Woods, have been observed to be less crowded and better able to preserve the purity of the surroundings than nearby public parks.

Private property rights might also spur biodiversity. In an article in *Cato Journal*, David Schap, a professor at Holy Cross University, and Andrew T. Young, a professor at Emory University, provide several examples of how this might occur. For instance, game ranching "is the private ownership of wildlife carried out on private property, typically for profit." Because state and local governments own domestic wildlife, game ranchers are free to raise exotic breeds. In many cases, these "exotics" are threatened or endangered species that would not be raised on traditional farms. Schap and Young conclude: "In the case of game ranching, profit motivates ranchers to cultivate a product—biodiversity of game—as an investment, the cost of which is less than hunters value the chance at exotic sport, pelts, and meat. The result: biodiversity is provided by the exchange of property rights from ranchers to hunters."

Private property rights may not be wholly compatible with the needs of the environment. Opponents to free-market environmentalism assert that property rights advocates ignore the positive effects of government regulations and contend that not all property owners are interested in preserving natural beauty. As it stands, middle ground may never be found between those who fear the economic consequences of environmental regulations and those who think that profit-making precludes the protection of wildlife and open spaces. In the following chapter the cartoonists explore the conflict between economy and ecology.

Examining Cartoon 1:
"I'm in the Information Economy"

About the Cartoon

Many people believe that as the U.S. economy moves away from one that is centered around traditional industries to one that is based on computers and technology, the environment will improve because modern industries do not produce smoke or other causes of air pollution. However, as Chip Bok suggests in the accompanying cartoon, that belief is misguided. Computers require electricity to run, which is indicated by the man shoveling coal into the machine labeled "power supply." While the man at the computer claims he is not part of the "smokestack economy," studies have shown that computers, the Internet, and other telecommunication equipment consume between 2 and 8 percent of all U.S. electricity, a number that will likely rise as the "information economy" develops.

About the Cartoonist

Chip Bok, or Arthur B. Bok III, is the editorial cartoonist for the *Akron (Ohio) Beacon Journal*. His cartoons have appeared in more than one hundred publications, including the *Washington Post*, the *New York Times*, *Newsweek*, and *Reason*. Among the first place awards he has garnered for his work are the 1995 National Cartoonist Society award for best editorial cartoonist, the 1993 Berryman Award given by the National Press Foundation for editorial cartoons, and the 1993 H.L. Mencken Award for editorial cartooning from the Free Press Association.

Examining Cartoon 2:
"Protecting Our Really Important Species"

About the Cartoon

In May 2003 Christie Todd Whitman resigned as administrator of the Environmental Protection Agency. Whitman was a moderate whose views on global warming, pollution controls, and other important environmental issues were often at odds with the more conservative George W. Bush administration.

In this cartoon Ben Sargent sarcastically suggests that Whitman's resignation will allow the Bush administration—represented in his drawing by the White House—to protect "really important species." According to Sargent, with Whitman out of the picture, the White House will reveal its true agenda: protecting business interests instead of preserving the environment. He depicts these economic interests, including profit and exploitation, as hybrids of humans and four animals, including vultures, snakes, and lizards. Bush nominated Utah governor Mike Leavitt to replace Whitman. The nomination was confirmed in November 2003; environmental groups criticized the choice, arguing that Leavitt has given millions of acres to developers and inadequately protected Utah's environment.

About the Cartoonist

Ben Sargent is the editorial cartoonist for the *Austin American-Statesman* in Texas. He won the 1982 Pulitzer Prize for editorial cartooning and is a former president of the Association of American Editorial Cartoonists. Sargent is also the author of *Texas Statehouse Blues* and *Big Brother Blues*.

Examining Cartoon 3:

"I Just Can't Figure Out How They Get Away with It"

About the Cartoon

In this cartoon Bob Lang explores the idea that industries, in this case the agricultural industry, are blamed for pollution while other sources of contamination are ignored. He depicts a protester insulting the operators of a hog farm, calling them "antienvironmentalists." However, the hog farm is a zero discharge facility, meaning that no waste products enter the water supply. While the farmers are being wrongly blamed for polluting the water, the protester ignores the harmful chemicals, phosphates and nitrates, which are released by the city's sewage system. Phosphates and nitrates are harmful to the environment because they encourage algae growth. When the algae dies, bacteria is released as part of the decomposition process. The oxygen used by the bacteria during the process is then unavailable to the fish in that water, leading to higher mortality rates. Lang is thus concluding that industries are better able to protect the environment than governments are.

About the Cartoonist

Bob Lang is a political cartoonist whose works have appeared in *Insight* magazine, the *Washington Times*, and the *Washington Post*. He also contributes cartoons to several websites, including CNSNews.com and CNN All Politics, and is a two-time winner of the National Newspaper Publishers' Best Editorial Cartoon of the Year award.

Chapter 3

The Global Warming Debate

Preface

A widely held belief of environmentalists and scientists is that Earth's temperature has been rising steadily over the past century, a phenomenon known as global warming. Research has suggested that global warming is caused by the increasing amount of carbon dioxide and other greenhouse gases in the atmosphere, an increase that purportedly results from activities such as deforestation and consumption of fossil fuels. If the research is accurate, global warming could have serious ecological consequences, from rising sea levels to extreme weather conditions. For example, global warming may have been responsible for a deadly heat wave in Europe in summer 2003, in which triple-digit temperatures (rarely seen on the Continent) were blamed for thousands of deaths.

However, while extreme weather is one potential consequence of Earth's changing climate, many people believe that if global warming does exist, its effects might not be solely ecological. Some studies have suggested that higher temperatures could make wider swaths of Earth's population vulnerable to fatal diseases.

Insect-borne and waterborne diseases may spread as a result of increased global temperatures. Paul Kingsnorth, a deputy editor of *Ecologist* magazine, writes, "[Virtually] all experts seem to agree that one effect of climate change will be to increase the range of the malarial mosquito," adding that by the second half of this century as many as 80 million more cases could occur each year. The affected areas will be in North America and Europe, regions that had previously not been warm enough to attract those mosquitoes. Other studies have found that the rate of tick-borne encephalitis has tripled in parts of northern Europe. Citing a World Health Organization report that claims heightened temperatures may reduce water supply and thus decrease the availability of water for cleaning and sanitation, Kingsnorth also observes that cholera and other waterborne diseases will become more virulent.

Even if the number of fatalities from these diseases is limited, the spread of infectious illnesses can have serious consequences. Paul R. Epstein, the associate director of the Center for Health and the Global Environment at Harvard Medical School, explored this issue in an article for *Scientific American*. Epstein argues,

> As the atmosphere has warmed over the past century, droughts in arid areas have persisted longer, and massive bursts of precipitation have become more common. Aside from causing death by drowning or starvation, these disasters promote by various means the emergence, resurgence and spread of infectious disease. That prospect is deeply troubling, because infectious illness is a genie that can be very hard to put back into its bottle. It may kill fewer people in one fell swoop than a raging flood or an extended drought, but once it takes root in a community, it often defies eradication and can invade other areas.

Numerous researchers who dispute the findings of these studies believe either that global warming does not exist or that it is not a serious threat to Earth and its inhabitants. For the past decade, few environmental issues have been as controversial as global warming. The cartoonists in the following chapter explore the global warming debate. If temperatures have been rising, then finding a solution that protects the health of the environment and stymies the spread of infectious diseases will be crucial.

Examining Cartoon 1:

"Global Warming Is Simply One of Nature's Quirky Cycles"

About the Cartoon

In this cartoon Don Wright pokes fun at people who believe global warming is not a problem. He depicts a man who claims that global warming is nothing more than a cyclical weather pattern; meanwhile, the man is shown being gradually submerged in a rising ocean. According to research on global warming, one of the effects of the increase in global temperatures—which could increase by as much as six degrees over the next century—is a rise in the sea level. A study by a scientist at the Environmental Defense Fund found that the sea level will rise by a meter during the twenty-first century. This increase will be the result of oceans expanding as they warm and the melting of ice caps and glaciers. While Wright presents this possibility comically, a sharp change in the level of seawater could have a devastating effect on the world, particularly on small island nations that will be affected by increased coastal erosion and the subsequent economic and social consequences.

About the Cartoonist

Don Wright is the editorial cartoonist of the *Palm Beach Post* and a two-time winner of the Pulitzer Prize.

Examining Cartoon 2:
"End Global Warming"

About the Cartoon

Few environmental debates are more controversial than the one over the existence and extent of global warming. Scott Stantis casts doubt on the threat of global warming in this cartoon. The drawing depicts two men clad in winter clothes who are unable to operate a rotating sign reading "End Global Warming." One of the men notes that the sign is frozen. With that image, Stantis is mocking the claim that the temperature of the earth is rising. The irony is heightened by the fact that the sign is solar powered. His cartoon echoes the findings of scientists and researchers who report that while there was warming from the mid–nineteenth century until the 1940s, weather satellite data have shown no proof of heightened temperatures since that time.

About the Cartoonist

Scott Stantis is the editorial cartoonist of the *Birmingham News*. His cartoons are syndicated to more than five hundred newspapers. In addition to his editorial work, he draws the comic strip *The Buckets*.

Examining Cartoon 3:
"You're Kidding, Right?"

About the Cartoon

In December 1997 more than 150 countries negotiated the Kyoto Protocol, a United Nations treaty aimed at reducing carbon dioxide emissions in the hopes that such an undertaking would reduce the problem of global warming. President George W. Bush withdrew U.S. support for the treaty following his inauguration in 2001. In this cartoon Michael Ramirez supports Bush's decision, suggesting that implementation of the protocol would be suicidal to American interests. Ramirez alludes to the fact that the treaty was negotiated in Japan by showing Uncle Sam sitting in a teahouse and speaking with disbelief when he sees a sword reading "Kyoto Protocol" and a scroll informing him to slit his abdomen.

In Japanese culture, disgraced officials were once expected to kill themselves with a sword, an act known as hara-kiri. Ramirez's Uncle Sam cannot believe he is expected to follow through with the command.

According to opponents of the Kyoto Protocol, the treaty would have devastated the U.S. economy. Had the United States agreed to the protocol, it would have had until 2012 to reduce its carbon emissions to 7 percent below 1990 levels. The Department of Energy had estimated that agreeing to the treaty would have led to a nearly $400 billion decrease in the gross national product by 2010 while raising the cost of electricity by 86.4 percent. Other concerns raised by opponents of the protocols were that oil and gas prices could rise, while American companies might choose to move to developing nations that are not beholden to the restrictions; such acts would have caused significant job losses in the United States.

About the Cartoonist

Michael Ramirez is the editorial cartoonist of the *Los Angeles Times* and the recipient of the 1994 Pulitzer Prize for editorial cartooning.

Examining Cartoon 4:
"You'll Have To Cut Your Emissions"

About the Cartoon

The Kyoto Protocol was a United Nations effort aimed at mitigating the effects of global warming. More than one hundred fifty nations negotiated the treaty in December 1997, which included the provision that by 2012 the United States would reduce its carbon dioxide emissions to 7 percent below 1990 levels. Many opponents of the treaty felt this demand placed too great a burden on the United States, charging that such a reduction would raise energy costs and could lead to job losses if companies move out of America. One point of rancor was the fact that developing nations such

as China, India, and Mexico would be exempt from emission-reduction
requirements.

Kirk Anderson questions the claim that Third World nations should play a bigger role in emissions reduction in this cartoon. His cartoon suggests that such a demand is hypocritical. A stereotypical American (wearing a hat with a flag motif) is shown driving a carbon dioxide–spewing truck full of polluting factories to Kyoto. He tells the Third World, represented by a cyclist smoking a cigarette, that it needs to cut down on its emissions, even though the Third World's effect on the environment clearly pales in comparison to that of the United States. Despite the argument presented in this cartoon, which was also stated by key environmental organizations, President George W. Bush ordered the United States to withdraw from the protocol in 2001.

About the Cartoonist

Kirk Anderson is the editorial cartoonist of the *Toledo Blade*. His cartoons have also appeared in hundreds of magazines, newspapers, and books, most notably the *New York Times*, *Los Angeles Times*, and *Washington Post*. Anderson also teaches editorial cartooning to a variety of individuals and groups.

Chapter 4

Protecting the Environment

Preface

The United States contains 6 percent of Earth's population but consumes nearly 30 percent of the available natural resources. Because it uses such a disproportionate share of resources, the United States is often expected to lead the way in finding ways to protect the environment. However, neither the United States, nor any other developed nation, can be expected to bear the full burden of Earth's ecological problems. Many environmental scholars assert that the third world should also help find ways to guard natural resources and end environmental ills.

Environmental problems in developing nations can be linked to several causes, including poverty and misuse of natural resources. Without adequate funds, third world governments cannot build sanitation facilities that will provide their residents with clean water. Food supply and safety also suffer in nations with serious ecological problems. Norway's minister of foreign affairs, Thorbjørn Jagland, points out in an article published by the United Nations Environmental Programme that "a healthy environment for food production is essential for a sustainable food supply and good nutrition." Such an environment requires efficient use of resources, soil conservation, and the protection of genetic diversity.

However, as Peter Huber, a research fellow at the Manhattan Institute for Policy Research, explains, "Poor countries are horribly bad at conservation because they lack the capital and know-how that we have put to such good use. . . . They consume little, but they are wasteful and destructive." For example, South American farmers cut down an unnecessary amount of trees in tropical rain forests because they need more land on which to farm and lack knowledge of more productive and environmentally sound agricultural techniques.

Huber is among those who believe capitalism and wealth will enable third world nations to better improve the environment (one study has found that levels of air and water pollution begin to significantly decline once per capita annual income reaches at least twenty-five hundred dollars).

Not all environmental writers believe that developing nations will benefit from first world examples. Instead, some argue that third world nations can better the environment through the use of indigenous solutions. Examples include vegetarianism in India and traditional agricultural methods that do not rely on pesticides and therefore do not contribute to groundwater pollution. Neeru Singh, the executive officer at the United Nations Centre for Human Settlement, suggests combining traditional solutions with modern science to solve India's lack of clean drinking water. Singh recommends rainwater harvesting, in which rainwater is stored in reservoirs or underground rivers so that it is accessible when needed. According to Singh, "The advantages of traditional methods such as rain harvesting are numerous. . . . They can give high agricultural returns and their installation and maintenance are cost-effective. They are also highly sustainable."

Whether through first world or third world solutions, the world must find ways to protect the environment. In this chapter the cartoonists evaluate several responses to modern ecological problems.

Examining Cartoon 1:
"Now We Can Relax and Breathe Easier!"

About the Cartoon

Monte Wolverton criticizes the actions of the Environmental Protection Agency (EPA) in this cartoon. In particular he targets the EPA's decision in November 2002 to relax regulations governing industrial pollution. Under the revised regulations, a company with multiple sources of air pollution (for example, several smokestacks) can clean up one emissions source and allow the remaining sources to become dirtier, as long as the company's total amount of emissions declines. In another change from previous regulations, factories can make significant alterations without being required to install pollution reduction equipment. Environmentalists expressed concern that the EPA's decision will worsen air pollution. In his cartoon Wolverton mockingly depicts industrial polluters wearing gas masks in front of smoke-spewing factories, claiming that they can "breathe easier" in the wake of the relaxed laws. While the industrialists can now operate their factories with greater ease, the environment around them is becoming increasingly unsafe.

About the Cartoonist

Monte Wolverton is a cartoonist for *MAD* magazine and the managing editor of the magazine *Plain Truth*.

Examining Cartoon 2:
"The Road to Hell . . . Is Paved!"

About the Cartoon

Dan Wasserman contends in his cartoon that cutting back on driving would reduce pollution, end the suburban sprawl that has led to the destruction of wildlife habitats and the loss of open space, and otherwise improve the quality of the environment. However, his cartoon suggests that the highway lobby—the oil, automobile, tire, and cement industries—opposes such efforts. The lobbyist in his cartoon offers a variation of the aphorism "the road to hell is paved with good intentions," instead declaring simply that "the road to hell is paved"—an idea that appears to delight the lobbyist. The highway lobby has been accused of thwarting efforts to reduce air pollution and the other problems caused by automobiles. For instance the highway lobby has led efforts to weaken the Clean Air Act, such as encouraging the passage of legislation that allows highway construction projects to continue even if they fail to comply with state clean-air improvement plans. Consequently, Wasserman's cartoon suggests, the highway lobby is at odds with the environmental needs and wishes of many Americans.

About the Cartoonist

In addition to his job as editorial cartoonist for the *Boston Globe*, Dan Wasserman has published *We've Been Framed*, a collection of cartoons about the Reagan presidency. His cartoons have also been published in *Time* and other national magazines.

Examining Cartoon 3:
"Thinning Out the Forest"

About the Cartoon

In this cartoon Dick Wright offers a sarcastic comparison of two approaches to protecting forests. In the "A" panel, he shows a forest that has been decimated by fire, its few remaining trees little more than trunks and leafless branches. By comparison, the forest in the "B" panel includes a few trees that have been cut down, but the remaining trees are healthy. According to Wright, the "A" forest is the method preferred by the environmental organization Sierra Club. The Sierra Club has stated its opposition to cutting down trees while also asserting that prescribed burning can restore minerals to the soil while also creating habitats for fish, birds, and animals. Wright disagrees with this notion and suggests that selective logging is the saner solution to preserving forests.

About the Cartoonist

Dick Wright is the political cartoonist of the *Columbus Dispatch* in Ohio. His cartoons are syndicated to more than four hundred newspapers and have also been published in *Time*, *Newsweek*, and *U.S. News & World Report*. He was a Pulitzer Prize finalist in 1983.

Organizations to Contact

The editors have compiled the following list of organizations concerned with the issues debated in this book. The descriptions are derived from materials provided by the organizations. All have publications or information available for interested readers. This list was compiled on the date of publication of the present volume; the information provided here may change. Be aware that many organizations take several weeks or longer to respond to inquiries, so allow as much time as possible.

American Council on Science and Health (ACSH)
1995 Broadway, 2nd Floor, New York, NY 10023-5860
(212) 362-7044 • fax: (212) 362-4919
e-mail: acsh@acsh.org • website: www.acsh.org

ACSH is a consumer education consortium concerned with environmental and health-related issues. The council publishes the quarterly *Priorities*, position papers such as "Global Climate Change and Human Health," and numerous reports, including *Arsenic, Drinking Water, and Health* and *The DDT Ban Turns 30*.

Canadian Centre for Pollution Prevention (C2P2)
100 Charlotte St., Sarnia, ON, N7T 4R2 Canada
(800) 667-9790 • fax: (519) 337-3486
e-mail: info@c2p2online.com • website: www.c2p2online.com

The Canadian Centre for Pollution Prevention is Canada's leading resource on ways to end pollution. It provides access to national and international information on pollution and prevention, online forums, and publications, including *Practical Pollution Training Guide* and the newsletter *at the source*, which C2P2 publishes three times a year.

Cato Institute
1000 Massachusetts Ave. NW, Washington, DC 20001-5403
(202) 842-0200 • fax: (202) 842-3490
e-mail: cato@cato.org • website: www.cato.org

The Cato Institute is a libertarian public policy research foundation that aims to limit the role of government and protect civil liberties. The institute believes EPA regulations are too stringent. Publications offered on the website include the bimonthly *Cato Policy Report*, the quarterly journal *Regulation*, the paper "The EPA's Clear Air-ogance," and the book *Climate of Fear: Why We Shouldn't Worry About Global Warming*.

Competitive Enterprise Institute (CEI)
1001 Connecticut Ave. NW, Suite 1250, Washington, DC 20036
(202) 331-1010 • fax: (202) 331-0640
e-mail: info@cei.org • website: www.cei.org

CEI is a nonprofit public policy organization dedicated to the principles of free enterprise and limited government. The institute believes private incentives and property rights, rather than government regulations, are the best way to protect the environment. CEI's publications include the newsletter *Monthly Planet* (formerly *CEI Update*), *On Point* policy briefs, and the books *Global Warming and Other Eco-Myths* and *The True State of the Planet*.

Environmental Protection Agency (EPA)
Ariel Rios Building, 1200 Pennsylvania Ave., NW
Washington, DC 20460
(202) 272-0167
website: www.epa.gov

The EPA is the federal agency in charge of protecting the environment and controlling pollution. The agency works toward these goals by enacting and enforcing regulations, identifying and fining polluters, assisting businesses and local environmental agencies, and cleaning up polluted sites. The EPA publishes periodic reports and the monthly *EPA Activities Update*.

Environment Canada

351 St. Joseph Blvd., Gatineau, Quebec K1A 0H3 Canada
(819) 997-2800 or (800) 668-6767
fax: (819) 953-2225 • TTY: (819) 994-0736
e-mail: enviroinfo@ec.gc.ca • website: www.ec.gc.ca

Environment Canada is a department of the Canadian government. Its goal is the achievement of sustainable development in Canada through conservation and environmental protection. The department publishes reports, including *Environmental Signals 2003*, and fact sheets on a number of topics, such as acid rain and pollution prevention.

Foundation for Clean Air Progress (FCAP)

1801 K St. NW, Suite 1000L, Washington, DC 20036
(800) 272-1604
e-mail: info@cleanairprogress.org
website: www.cleanairprogress.org

FCAP is a nonprofit organization that believes the public is unaware of the progress that has been made in reducing air pollution. The foundation represents various sectors of business and industry in providing information to the public about improving air quality trends. FCAP publishes reports and studies demonstrating that air pollution is on the decline, including *Breathing Easier About Energy —A Healthy Economy and Healthier Air* and *Study on Air Quality Trends, 1970-2015*.

Global Warming International Center (GWIC)

22W381 75th St., Naperville, IL 60565
(630) 910-1551 • fax: (630) 910-1561
website: www.globalwarming.net

GWIC is an international body that provides information on global warming science and policy to industries and governmental and nongovernmental organizations. The center sponsors research supporting the understanding of global warming and ways to reduce the problem. It publishes the quarterly newsletter *World Resource Review*.

National Resources Defense Council (NRDC)
40 W. 20th St., New York, NY 10011
(212) 727-2700 • fax: (212) 727-1773
e-mail: nrdcinfo@nrdc.org • website: www.nrdc.org

The NRDC is a nonprofit organization with more than four hundred thousand members. It uses laws and science to protect the environment, including wildlife and wild places. NRDC publishes the quarterly magazine *OnEarth* (formerly *Amicus Journal*) and hundreds of reports, including *Development and Dollars* and the annual report *Testing the Waters*.

Pew Center on Global Climate Change
2101 Wilson Blvd., Suite 550, Arlington, VA 22201
(703) 516-4146 • fax: (703) 841-1422
website: www.pewclimate.org

The Pew Center is a nonpartisan organization dedicated to educating the public and policy makers about the causes and potential consequences of global climate change and informing them of ways to reduce the emissions of greenhouse gases. Its reports include *Designing a Climate-Friendly Energy Policy* and *The Science of Climate Change*.

Political Economy Research Center (PERC)
2048 Analysis Dr., Suite A, Bozeman, MT 59718
(406) 587-9591
e-mail: perc@perc.org • website: www.perc.org

PERC is a nonprofit research and educational organization that seeks market-oriented solutions to environmental problems. The center holds a variety of conferences and provides environmental educational material. It publishes the quarterly newsletter PERC Reports, commentaries, research studies, and policy papers, among them "Economic Growth and the State of Humanity" and "The National Forests: For Whom and For What?"

Sierra Club
85 Second St., Second Floor, San Francisco, CA 94105-3441
(415) 977-5500 • fax: (415) 977-5799
e-mail: information@sierraclub.org • website: www.sierraclub.org

The Sierra Club is a grassroots organization with chapters in every state that promotes the protection and conservation of natural resources. The organization maintains separate committees on air quality, global environment, and solid waste, among other environmental concerns, to help achieve its goals. It publishes books, fact sheets, the bimonthly magazine *Sierra*, and the *Planet* newsletter, which appears several times a year.

Union of Concerned Scientists (UCS)
2 Brattle Square, Cambridge, MA 02238
(617) 547-5552 • fax: (617) 864-9405
e-mail: ucs@ucsusa.org • website: www.ucsusa.org

UCS aims to advance responsible public policy in areas where science and technology play important roles. Its programs emphasize transportation reform, arms control, safe and renewable energy technologies, and sustainable agriculture. UCS publications include the twice-yearly magazine *Catalyst*, the quarterly newsletter *earthwise*, and the reports *Greener SUVs* and *Greenhouse Crisis: The American Response*.

Worldwatch Institute
1776 Massachusetts Ave. NW, Washington, DC 20036-1904
(202) 452-1999 • fax: (202) 296-7365
e-mail: worldwatch@worldwatch.org
website: www.worldwatch.org

Worldwatch is a nonprofit public policy research organization dedicated to informing the public and policy makers about emerging global problems and trends and the complex links between the environment and the world economy. Its publications include *Vital Signs*, issued every year, the bimonthly magazine *World Watch*, the Environmental Alert series, and numerous policy papers, including "Unnatural Disasters" and "City Limits: Putting the Brakes on Sprawl."

For Further Research

Books

Ronald Bailey, ed., *Earth Report 2000: Revisiting the True State of the Planet*. New York: McGraw-Hill, 2000.

John J. Berger, *Beating the Heat: Why and How We Must Combat Global Warming*. Berkeley, CA: Berkeley Hills Books, 2000.

Daniel Botkin et al., *Forces of Change: A New View of Nature*. Washington, DC: National Geographic Society, 2000.

Pamela S. Chasek, ed., *The Global Environment in the Twenty-First Century: Prospects for International Cooperation*. Tokyo: United Nations University Press, 2000.

Jack Doyle, *Taken for a Ride: Detroit's Big Three and the Politics of Pollution*. New York: Four Walls Eight Windows, 2000.

Simon Dresner, *The Principles of Sustainability*. London: Earthscan, 2002.

Richard Ellis, *The Empty Ocean: Plundering the World's Marine Life*. Washington, DC: Island Press, 2003.

Hilary French, *Vanishing Borders: Protecting the Planet in the Age of Globalization*. New York: W.W. Norton, 2000.

Ron Fridell, *Global Warming*. New York: Franklin Watts, 2002.

Michael H. Glantz, *Climate Affairs: A Primer*. Washington, DC: Island Press, 2003.

Indur Goklany, *Clearing the Air: The Real Story of the War on Air Pollution*. Washington, DC: Cato Institute, 1999.

Joyeeta Gupta, *Our Simmering Planet: What to Do About Global Warming?* London: Zed Books, 2001.

Peter Huber, *Hard Green: Saving the Environment from the Environmentalists: A Conservative Manifesto*. New York: Basic Books, 1999.

Robert Hunter, *Thermageddon: Countdown to 2030*. New York: Arcade, 2003.

Andrew Jamison, *The Making of Green Knowledge: Environmental Politics and Cultural Transformation*. Cambridge, UK: Cambridge University Press, 2001.

Bjørn Lomborg, *The Skeptical Environmentalist: Measuring the Real State of the World*. Cambridge, UK: Cambridge University Press, 2001.

Patrick J. Michaels and Robert C. Balling Jr., *The Satanic Gases: Clearing the Air About Global Warming*. Washington, DC: Cato Institute, 2000.

Robert L. Nadeau, *The Wealth of Nature: How Mainstream Economics Has Failed the Environment*. New York: Columbia University Press, 2002.

Shannon C. Petersen, *Acting for Endangered Species: The Statutory Ark*. Lawrence: University Press of Kansas, 2002.

Richard C. Porter, *The Economics of Waste*. Washington, DC: Resources for the Future, 2002.

Kent E. Portney, *Taking Sustainable Cities Seriously: Economic Development, the Environment, and Quality of Life in American Cities*. Cambridge, MA: MIT Press, 2003.

Richard P. Reading and Brian Miller, eds., *Endangered Animals: A Reference Guide to Conflicting Issues*. Westport, CT: Greenwood Press, 2000.

Philip Shabecoff, *Earth Rising: American Environmentalism in the 21st Century*. Washington, DC: Island Press, 2000.

Vandana Shiva, *Water Wars: Privatization, Pollution and Profit*. Cambridge, MA: South End Press, 2002.

Lawrence Slobodkin, *A Citizen's Guide to Ecology*. Oxford, UK: Oxford University Press, 2003.

Eric R.A.N. Smith, *Energy, the Environment, and Public Opinion*. Lanham, MD: Rowman & Littlefield, 2002.

Barbara Taylor, *How to Save the Planet*. New York: Franklin Watts, 2001.

Edward O. Wilson, *The Future of Life*. New York: Alfred A. Knopf, 2002.

Periodicals

Ronald Bailey, "Earth Day, Then and Now," *Reason*, May 2000.

Lester R. Brown, "An Economy for the Earth," *Humanist*, May/June 2002.

Thomas J. Crowley, "Causes of Climate Change over the Past 1000 Years," *Science*, July 14, 2000.

Gregg Easterbrook, "Get the Easy Greenhouse Gases First," *New York Times*, August 29, 2000.

Economist, "Out of the Blue," November 2, 2002.

Ian Frazier, "As the World Burns," *Mother Jones*, March/April 2003.

Ross Gelbspan, "Reality Check," *E: The Environmental Magazine*, September/October 2000.

Llewellyn D. Howell, "Global Warming, Global Warning," *USA Today*, March 2000.

Peter Huber, "Wealth Is Not the Enemy of the Environment," *Vital Speeches of the Day*, April 1, 2000.

Matt Kaplan, "Plight of the Condor," *New Scientist*, October 5, 2002.

Jane Holtz Kay, "Infernal Combustion," *In These Times*, August 8, 1999.

Lisa Mastny and Hilary French, "Crimes of (a) Global Nature," *World Watch*, September/October 2002.

Bill McKibben, "Too Hot to Handle," *New York Times*, January 5, 2001.

George Melloan, "Scrapping Kyoto May Prove to Be Bush's Finest Act," *Wall Street Journal*, April 3, 2001.

J. Madeleine Nash, "Fireproofing the Forests," *Time*, August 18, 2003.

John Pike, "Good News Is That Bad News Is Wrong," *Insight on the News*, October 1, 2002.

April Reese, "Bad Air Days," *E: The Environmental Magazine*, November/December 1999.

Payal Sampat, "The Hidden Threat of Groundwater Pollution," *USA Today Magazine*, July 2001.

Roddy Scheer, "Parks as Lungs," *E: The Environmental Magazine*, November/December 2001.

Gretel H. Schueller, "Wasting Away," *OnEarth*, Fall 2002.

Todd Seavey, "The Killer Fog," *Reason*, April 2003.

Jane S. Shaw, "Private Property Rights, Not Ideologies, Are the Crux," *Independent Review*, Summer 2002.

S. Fred Singer, "Cool Planet, Hot Politics," *American Outlook*, Summer 2000.

William K. Stevens, "Global Warming: The Contrarian View," *New York Times*, February 29, 2000.

Robert T. Watson and David Wojick, "Do Scientists Have Compelling Evidence of Global Warming?" *Insight on the News*, March 12, 2001.

Index